Journey to Asian Leadership

Sunil Kankal

DEDICATION

I dedicate this book "Journey to Asian Leadership" to my wife Jyoti and my Son Rohit. Thank you for all your endless love, unconditional support and encouragement.
I love you.

CONTENTS

Chapters	Page No

ACKNOWLEDGMENTS

First and Foremost, I have to thank my parents for their love and unconditional support in my life. Thank you both for giving me strength to chase my dreams. My sisters, brother and my cousins deserve my wholehearted thanks as well.

I would like to acknowledge and express my gratitude to Mr Soloman Isaac, Mr. Shivaji Godse and Mr. Chetan choudhary for their guidance and magnificent support in developing courage and confidence in this journey.

To, all my friends, thank you for your understanding and encouragement in my many many moments of crises. Your friendship makes my life a wonderful experience.

Thank you, God for always being there for me.

1 INTRODUCTION

In today's Technology driven generation, larger Paradigm shifts are taking place thereby shortening the decade timeline, and we have great minds to rationalize anything that comes into our mind.

I am sharing my Journey I did from System Integrator to CTO. It has been very common amongst everyone that, what all the contribution you do for the company, driven by your Internal emotion will exceed Stakeholders Expectation by fostering Innovation at the work place.

The first Career Challenge got me to Design the Digital Data Logger Product for the measurement of Furnace temperature. I applied the Geometry Analytics principle to design the Prototype successfully which got into manufacturing. The result bought so much excitement to owner of the company who clicked multiple photographs as if he is kissing the product. It shows his delight and happiness in the project.

Progressively I transitioned ahead to Rollout Mobile Data centre on the Wire-line communication Infrastructure at 7 new offshore Oil Platforms and then with IT Hardware and Networking Vertical, where I conducted Boot Camp on ISO 9000 Quality Training to 35 Franchisee units resulting 30% profit growth.

During the Dot com burst, I spearheaded challenge to setup Technology Platform for Startup ITeS within 40m USD budget. I formed and grew 400+ Team, developed skills into Vendor Governance, IT consolidation and delivered 42% CAGR savings. I added my skills to build Global Data centre with Disaster Recovery mechanism on cloud and added paradigm shift from person to process oriented operations on Auto Pilot.

In 2008, I got on board with US based MNC for IP Telephony software and then moved to 500m$ FCMG group to implement one number calling platform on Siemens Fixed Mobile Convergence. I was able to deliver gross savings of 42% on CAGR from IT consolidation to carbon credits.

When I effectively crossed the board in 2008, I knew the group as a client & partner for proposing the interesting challenge in Asia, but not really the company from inside. I also must admit that it was difficult to leave my previous company and more importantly it was a fair jump.

I moved to Leadership position with International Business Relationships and Technology skills from fundamental to functional level, FRU to CRM Integration, and Physical to application layer. I demonstrated flexibility with multi culture talents and tangible skills in business goal.

This consolidated experience from FMCG, ITeS & Offshore Industry groomed me to venture into Project Management Consulting.

I will discuss more on my Contribution that includes Large size Projects in Business Intelligence, Migration/Upgradation of Global Data centre over Wireless connectivity, IT Infrastructure and FSRRP project in UAE.

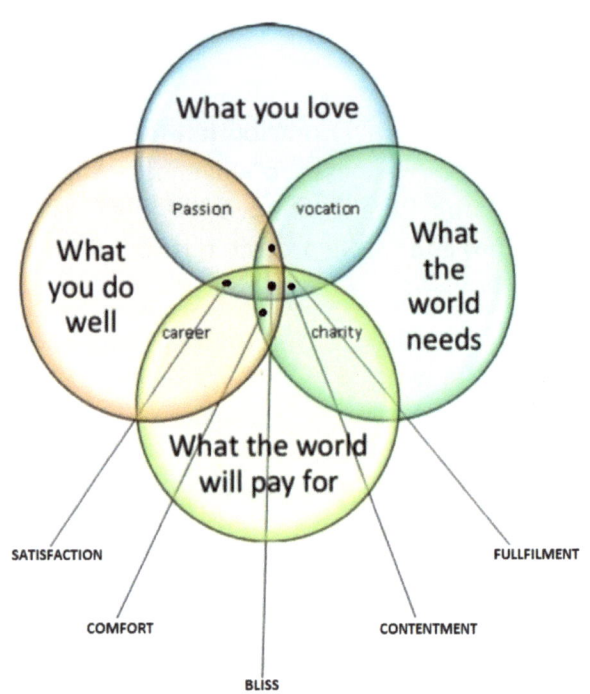

ART OF FULLFILMENT

2 TRANSFORMATION

Does the CIO becomes Chief Business Officer <CBO>? Will this be Real Impact to the Industry. The decisions taken by him / her will insulate towards ownership of the entire life cycle. Will this decision become source of his Inspiration? You may say this Chief Impact Officer / Chief Inspiration Officer or Chief Information Officer i.e C I O and some call it as OIC.(Oh I see). Is this Mirror effect?

In this technology driven business, CEO/CIO will add Impact with Inspiration & Innovation. And Technology is Enabler towards transformation.

We have so many examples where the transformation in the technology happened slowly in last 2 decades. During all these years, the Indian Culture got dormant to the US business style. The generation got inspiration more towards technology and it added more challenges relatively on the Cultural bonds. The community developed its own Niche methods of doing business with the control on the ownership. The Indian business is more of **Cultural Centric.** You will see that as business units become more reliant on the technology to do their work, the leaders are taking control over it.

As the Technology is getting Evolved towards Plug and Play mechanism, the support arm is inclined on the fast movements of Digitization and Technology importance gets rolled down to the Entertainment. More and more graduates started adding certification of IT from Hardware to Networking and these certifications got more attraction in magnetizing their career into IT.

This Transformation drives the IT Support Arm on the Certification driven community. This transformation is slowly bringing the decentralization in ownership of the Support System. Here are some of the Examples where the technology migrated Hosted platform to Cloud computing.

Amazon is offering you online process for publishing the book and offering ISBN code with Royalty payment. This is the strong example of Paradigm shift in the process, a real transformation.

Certification to be IT Enabler in business?

Right TALENT has Potential to drive the Invisible force and not the Certification driven force. Certifications are enabler to Change the Persons Attitude, so that some hidden Inspiration delivers the solution to certain extent.

Is this an Art of Fulfillment enabling Project Management or Service Delivery? You will find the answer. Now I would like to share simple Rules to become successful Project Manager in handling Agile Technology.

3 PROJECT MANAGEMENT

I had a small Networking Meet with the Industry Veteran. I introduced myself and the other side who seem to be a matured Industry veteran; started arguing on the topic of Project Management. What is that makes him to sit in the meeting. And immediately, he wants to call off the meeting with cup of coffee.

Here the significance is the confidentiality with the knowledge sharing. The cultural mechanism does not allow the person to open up and he is encapsulated with the fear.

I would like to provide you with some basic requirements here about the transformation in the Methodology. Some follow Agile method and other Waterfall approach. You will get better results by following

Project Management Methodology

1. Vaccinate Excusitis Disease
2. Improve Stickability in Task
3. Apply Single handling
4. Emotional Action
5. Appreciate & Contribute

Waterfall Approach

Frontline Executives are facing Challenges with the application development team that organizations still follow WATERFALL approach for legacy work and more interactive, iterative approach for new development work. They are facing barriers in applying the ITERATIVE approach due to Lack of business ownership.

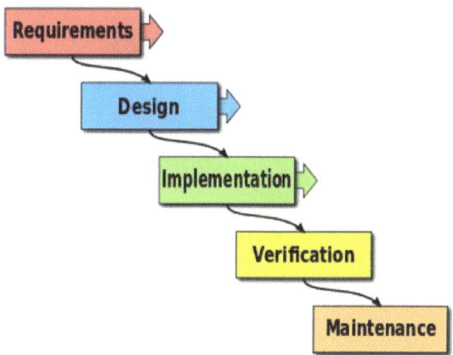

McKinsey Survey March 2014 comments "Executives from the business side, they are more than twice as likely to suggest replacing IT management as the best remedy."

IIT executives reports declining performance in relation to Managing IT Infrastructure and Governing IT Performance. Survey Insights shows "GROWING DISSATISFACTION WITH IT PERFORMANCE".

Changing IT leadership

Changing IT leadership became a priority to improve IT performance. The overall IT talent shortage is most pressing for analytics, but need will vary by sector. Companies will be forced to focus on culture and compensation.

Industry needs Leaders to bring Environment centric Execution tools which will not only address the Critical Business Gaps but streamline the Service Delivery, so called "Focused Attention". Have a better control on the Vendor Governance. Most of the Organization are Vendor driven with the addition of certification oriented skills.

IT Industry will require high level Strategy & tri-dimensional Analytics to support the Next Generation Infrastructure for C-Level decision makers.

With the Technology moving faster, the Leaders needs to Roll down Projects with mix of Waterfall and Agile methods for new development and inspire the team with the entire ownership of life cycle.

Tony Robbins say that **Invisible force of internal drive** activated is the most important thing in the world. All of us here have great minds. You know? Most of us here have great minds. Right I don't know about another category, but we all know how to think and with our minds we can rationalize anything and we can make anything happen. Because when **emotion** comes into it, the wiring changes in the way it functions. I believe that **Emotion is the force of life**. And so it's wonderful for us to think intellectually about the Project Management.

4 ULTIMATE RESOURCE

Who are the vendors? Is your company's Network, an SLA driven or Vendor driven?

What are the challenges that make them to contribute or complete the project?

I really want to know what's driving you and maybe invite you to explore where you are today, for two reasons.

One, so that you can contribute more in project management **And two**, so that hopefully we can not just understand other people more, but may be appreciate them more, and create the kinds of connections that can stop some of the challenges that we face in our society today. They're only going to get magnified by the very technology that's connecting us, because it's making us intersect.

I've had an obsession basically for 24 years, and that obsession has been, "What makes the difference in the quality, delivery and performance?"

How do you make a change? So that Performance is affected. What is it that's shaping that **person's ability to contribute, to do something beyond themselves,** So maybe the real question is, you know, I look at life and say, there's two master lessons.

One is: there's the <u>science of achievement</u>, which almost everything that's run is mastered to an amazing extent. That's "How do you take the invisible and make it visible," right? How do you take what you're dreaming of and make it happen? Whether it be your business, your contribution to society, money --whatever it is for you -- your body, your family.

But the other lesson of life that is rarely mastered is the Art of Fulfillment.

Because science is easy, right? We know the rules. You write the code. You follow and you get the results. Once you know the game you just do it, don't you? But **when it comes to fulfillment - that's an art**, and the reason is, it's about appreciation & contribution. You can only feel so much by yourself.

We **live in a therapy culture.** Most of us don't do that, but the culture's a therapy culture. And what I mean by that is the mindset that we copy and follow the result of others.

Do we want to make Biography as destiny. The past equals the future. And of course it does if you live there. But what we have to remind ourselves, you can know something intellectually, you can know what to do and then not use it, but not apply it.

So really, we're going to remind ourselves that "**Decision is the ultimate power**". That's what it really is. Now, when you ask people, you know, have you failed to succeed in the project or achieve something?

How many have ever failed to achieve something significant in your life? But if you ask people, why didn't you achieve something? Somebody who's working for you, you know, or a partner, or even yourself. When you fail to achieve a goal, what's the reason people say they fail to achieve? What do they tell you?

Didn't have the – knowledge.
Didn't have the – Proper Training.
Didn't have the – Time. I was busy with xyz.
Didn't have the – Information to complete.
Didn't have the – Email Contact.
Didn't have the Management to Resolve.
You know, I didn't have the right manager.
Didn't get the appraisal.

Where does the Dedication, determination, curiosity and resolution fly in mind? And what do all those, have in common?

They claim to you missing resources, and they may be accurate. You may not have the money; you may not have the Appraisal; but that is not the defining factor.

The **defining factor is never resources; it's RESOURCEFULNESS.** If you don't have the money, but you're creative and determined enough, you find the way. So **this is the ultimate resource.**

What determines your resources? If our **decisions shape destiny**, what determines it is three decisions. What are you going to focus on? Right now, you have to decide what you're going to focus on. In this second, consciously or unconsciously, the minute you decide to focus on something you've got to give it a meaning, and whatever that meaning is produces emotion. Is this the end or the beginning?

An emotion, then, creates what we're going to do or the action and this action is the emotional fitness, which is psychological strength. That's the difference in human beings that I've seen of the three million that I've been around. And **this is where fulfillment comes. Here is what you grow and contribute beyond the expectation.**

When I was in Kabul with the challenge to resolve the issue with the completion of 36 Financial/commercial Reports on Oracle Hyperion. I found that project was in Standby situation. The coding team was ocean miles apart from UAE Region. Client Manager wants to know the details of the code, and nothing moved for 18 months. And I took these two technical groups and did what I call an indirect negotiation. And the integration happened with the completion of project in 45 days with Pre-Launch of 25 Reports actually qualified by Senior Management. What actually happened is that the two groups not only came together, but they changed their beliefs and they worked together with the Launch of 80% Reports. This is real transformation in Project Management.

So my invitation to you is this: explore your web, the web in here -- the needs, the beliefs, the emotions that are controlling you, for two reasons: so there's more of you to give -- and achieve too, we all want to do it -- but I mean give, because that's what's going to fill you up. And secondly, so you can appreciate -- not just understand, that's intellectual, that's the mind -- but appreciate what's driving other people.

5 DID YOU KNOW

If you are one in a million in china, there are 1300 people just like you. The 25% of India's population with the Highest IQs…. Is greater than the total population of United States. India has more honors kids than America.

The top 10 in demand jobs for 2010… did not exist in 2004.

The US department of labor estimates that today's learner will have 10-14 jobs by age of 38. One in 4 workers has been with their current employer for less than a year and one in 2 has been there less than five years.

One out of 8 couples married in the US last year met online. There are 200 million registered users on MySpace. The No 1 ranked country in Broadband Penetration is BERMUDA, whereas #19 The United States and #22 Japan.

We are living in exponential times. There are 31 billion researches on Google every month. In 2007, the number was 2.7 billion.

Now to reach market audience of 50million

Radio reached in 38 years
TV in 13 years
Internet in 4 years
iPod in 3 years
And... Facebook 2 years
Number of Internet devices in 1994 was 1000
In 1992 it was 1000,000
In 2008 it is 1000,000,000

There are 540,000 words in the English language. It is estimated that 4 Exabyte's of unique information will be generated. That is more than previous 5000 years. The amount of Technical information is doubling.

NTT Japan has successfully tested a fiber optic cable that pushes 14 trillion bits/sec down a single strand of fiber. i.e 2660 CD or 210 million phone calls every sec. Data is tripling every six months and expected to do so for next 20 years.

The Predictions are China will soon become NUMBER ONE English Speaking Country in the World. China is Pioneer in transforming the Manufacturing of quality product at Most Economical price making it available for use to the weaker regions. This advantage empowers China to make product cycled to Use Once. We have examples where the transformation has impacted East Asia with its influence.

First and most important element of Change Pattern is very clearly seen on the transformation bought by the Technology into our lives. This transformation has moved the decade from 10 years to almost 3 years. Look at the Impact from the Linkedin Network and facebook. We say the world is small, but this has added great Impact to the

HR Industry and every HR is getting TONS of resume to either ATS system or the Resume bank. The Software Intelligence designed is not sufficient to screen the Resume and compare the traditional way of screening. It has made more difficult in searching right Talent.

Also it is discovered that this Impact of East Asia will bring Reorientation of the world from West to East. There will be Challenge in Sharing the Leadership from East to West and China will seek greater Incentives from this Integration. The growing influence will reshape the rules and institutions of International system. East Asia will feel that this rising Influence will bring power transition to add higher dimensions of Conflict and Cyber Security.

In terms of Telecom Leadership, Huawei is really ahead with the Rugged Technology and it is ruling the Mobile communication across East Asia and dormant in West.

Next Challenge is the Big Data growing at the exponential rate. We need to grow our minds in handling this Massive Data. Inter-operate and compare this data coming from different channels so that we feel satisfied in arriving to the perfect decision. Technocrats need to shift gears of Micro level thinking to Macro level thinking and at some point merge both of them, to bring the community satisfaction. The new challenge of Inspiration is required in coming next 10 years in Asia and China.

The Political Influence driving the Economics on the Technology as enabler in the Industry will add Impact to the need of matrix level multidimensional thinkers and Pyramid thinkers.

6 BEST PRACTICES

Only three things happens in organization Naturally
"friction, confusion and underperformance"
Everything else requires Leadership.
Peter Drucker

I will share some of the experience of Data centre, Business Intelligence and Operational Best Practices. It will provide direction in managing Big Data Analysis.

In 1994, I got training in Texas, USA to commission 7 No of Mobile Data Centre for the offshore Oil Industry, based on Wire-line communication Infrastructure on DEC 3800 MVax Platform. It was real excitement in the work to Identify and Isolate Cable faults, Calibration of Transducer/ Sensors / Photo Multiplier Tube and ensure 100% functioning with the Productivity Measuring & Recording Equipments. The Complete Unit is encapsulated in the

Racks with High Density Server, Spooling Tape Drive, Thermal Printer and System Interface Panel with SB Microcontroller to control Temperature, Pressure, Cable Tension, and Humidity. The complete System captures Digitized Data from Remote Measuring Instruments, displayed Real time graphic monitoring and stored Data on Spooled Tape drives. This knowledge gives similarity in today's Telecom Infrastructure and Project Management of Global Data centre.

In 2002-2007, set the vision, crafted strategy, built internal/external consensus, and restructured business plan for the Startup Unit pivoting to positive cash flow and profitability in < 180 days. Built a culture and team of talented, high impact professionals (15 No) backed with Vendor Support SLA across all functions.

I got the Robust Non-proprietary Technology Platform of 1000 seats with Hosted Data Center on ASPECT and DR site on Touchstar IP Telephony with Legacy Network on Enterasys, IBM Servers, Alcatel 7270, Tadiran PABX, Access control and Firewall Appliance. I added CCTV, Biometric Attendance, Fire Alarm System and UPS to Support Site Security. The Total budget in building up Infrastructure Platform with Remote manageability on Cloud got completed in 40m USD.

I developed skills of Vendor governance skills on customized SLA, Global AMC, MSA with clients and setting the Quality Standards on ISO 9001 & HIPPA. I demonstrated high Performance team Potential in Software Development, Quality control, Cost/Benefit Analysis for IT spending & Management Report and Best Practices in the Project Management.

BUSINESS INTELLIGENCE

I would like to share the Experience of Projects with Banking, BPO and Telecom Industry. How I applied skills in getting profits with use of Business Intelligence and Challenges faced with Big Data Analytics.

In 2012, a largest Telecom SP in UAE approached me on the challenge to deliver 36 Enterprise Reports on Oracle Hyperion. I found that Client A, an overseas Technical group, got all the coding on Waterfall approach, Client B Project Manager wants Agile method and nothing moved for 18 months. Probably this was Cultural gap all the time. After multiple meetings, I understood challenge in these two technical groups and did what I call an indirect negotiation with overseas Technical group for the Project Plan deliverables to expedite the ETL process and transition of 17m Customer Data from MS SQL to Oracle Hyperion.

And the integration happened with the Pre-Launch of 25 Reports actually qualified by Senior Management. What actually happened is that the two groups not only came together, but they changed their beliefs and they worked together with the Launch of 80% Reports. And more importantly is the level of discrimination the company and its people have encountered in trying to break the ice and make reputable business relations.

In 2011, Moved Profit fetch from 12% to 26% with 22 months ROI to the Investors Stake from 170000 Telecom lines. Here Client A did Investment of 1m USD with Client B supported with SLA. The Agreement said that the Return of Investment is from Monthly business bought by 170,000 Telecom circuits against yearly revenue collection of 2.2m USD. The Business Percentage did not moved above 12%.

The Challenge was to do Micro Level slice of Telecom Data of 172000 customers and derive the tangible Revenue growth above 25% revenue share.

I applied all the 3 dimensional report analysis skill and notified the correction at the CRM level, thereby moving the profit fetch to 26%.

What we did in brief is that 40% sites found underutilized with 60 ~ 80% Resources and noticed another 65% lines having average yield per line below normal.

With the BI Tools analysis, the revenue share increased from 14% to 26% and Got simplified with all the Revenue dependency on the customer usage, Escrow billing and other Technical factors. The ROI moved from 48 months to 18 months.

PRODUCTION ON CLOUD

Fundamentally Cloud Platform is evolution of Hosted technologies to BYOD and Pay per use model. As more and more business starting adding dependency on IT, the new transformation added more savings by way of Virtualization and consolidation at the level of Hardware and software licenses. Microcontroller boards got the place blade servers. The chipsets are more controlled by Shared Applications. This has opened Industry doors with BYOD to PPU Strategies.

During 2002-2007, I drove the team to add the new ideas in Managing the Operational Metrics, Performance, Productivity and automated the complete operations from Sale to verification, quality and closure on the Dashboard. This leveraged C-level approval on Client Payments and the Knowledgebase with Mobile Alerts for easy manageability of Level 1 problems by the onsite staff.

We got the development of Single Dashboard view to Manage IT operations, Asset Management, Historical Report and Real time MIS Reports. We encouraged team innovation in optimization of LAN/WAN, E1 IPLC, Internet Bandwidth, building VPN Network to deliver Secure Data for verification by client, Real time productivity tools for C-Level Executive. We were able to reduce the 3 day timeline of confirmed verification of sales to 30 min. The cloud impact not only improved the quality and delivery of sales, but also increased the cash flow with improvement in the client billing. This resulted in team motivation with high conversion of cross sales and benefit of incentives.

All-together we were able to arrive with the 60% cost savings from CAPEX, AMC and Profit of 25m USD Revenue from 102 Process of Telesales in Credit card, Telecom, Lead generation and Medical billing. Quantified consistent growth by 35% CAGR and operational yield of 85m$ with Performance driven Client delivery.

As a conclusion: The spirit of entrepreneurship is independent of the location! It means that wherever you are or you join an entrepreneur/ a family owned business or you create your own structure the KSF mentioned are Best Practices.

In 2008, I came on board as Executive Consultant for Denver USA based Company providing Software Solutions for BPO market. I worked on the Challenge to Implement the TELEPHONY Infra for the DIGITEX Client at South America in 15 days after Advance certification Program in IP Telephony on Web based Application suite with e-Superscripting feature on Centos / Asterix server.

In 2008-10, I got challenge in 500m$ FMCG group to UPGRADE 2500 user Legacy Telecom Switch. The Upgrade includes Implementation of Siemens FIXED MOBILE CONVERGENCE (FMC) on Unified communication Platform, Video conference and control wastage in the IT. The Role is to Set, Prioritize & Judiciously Justify the Capital Expenditure Request 'CER' to C-Level Matrix Management. The objective is to establish one number calling Strategy across the organization with 85+ locations. The Significance of FMC is to handle 4 digit calling between Wi-Fi and GSM Network over IP Network for the C-Level Team.

We got this Integration done on the SIP Interface with backbone of MPLS Network and GSM Gateway. FMC has leveraged business leaders with Mobility in one number calling. We faced new challenge with the non-availability of CAPEX for the Remote location and we were able to get this resolved from the Service provider by extending CUG on Micro BTS with Signal boosters over Leased Line Network.

We implemented concept of CARBON credits by pushing Class 5 switch, DLC Equipment on Panel cooler with Mobile alerts, Activity Sensor on PLCC technology & Remote manageability. We added Video conference at the Remote Plant on wireless radio and 9 Regional units.

The Wastage reduction Cycle is consolidation of PSTN billing, Printers, Video conference bridge at 9 location, and transitioning of BO Apps, Lotus Notes on the Linux OS delivered 42% Cost Savings to the organization.

In 2011, Got the Business Challenge form Media Industry having business into Secure Delivery of Movie to Digital Cinema Theatres, to implement Redundancy in the download of large size MPeG4 Files on the Multicast FTP application. We got this implemented on the Alvarion

WiMax product; Corinex PLCC and DSLAM switch to download multicast MPeG4 file.

In 2012, Executed E2E Deployment of 500 Bank ATM on V-Sat connectivity and Setting up Helpdesk operations in co-ordination with NCR/Diebold/Hughes Technical group for North India.

Internationally, I faced challenge to work on the FSRRP (Financial Security Rapid Response Project) where I Teamed with Europe based TagAttitude on Complete Integration of PM Cycle in MMP, ATM, PoS for Auto Payment System (APS) with 351 branches in 34 provinces governed by Central Bank – DAB in Kabul.

Next Internationally Managed LIVE Migration & Upgrade of Data centre on microwave Wireless Network for the Tier 1 Telecom Industry of 17m Customer Data within timeframe of 3 months. This resulted in Congestion free Network Performance. The Challenge faced here to implement Upgrade first and do the Migration of BDC site to new location. The Upgrade in the High Capacity SUN Server M9000, EMC Storage from DMax to VMax and Cisco Router was also important to reduce the Replication time of highly populated data built. I applied all the skills of Vendor Relations to the Principal technical groups and local team to manage the project within time.

In 2013, I managed the DC Site transitioning of 800 users ASPECT Platform within Timeframe of 72 hours downtime. We were able to switch ON all the Equipment from IBM Servers, EMC Storage on SAN in proper phased manner, well documented for startup of complete system at new location. We got all the background LAN cabling, CCTV, Desktop connectivity, Headsets and Telecom PSTN circuits ready to go live in 45 days.

7 VENDOR GOVERNANCE

Vendor is another family Element in the business. All the new technologies are introduced by him and the any missing functions can be outsourced for better quality.

An organization can follow proper Vendor Registration process with the details of stakeholders, Engineer strength and all the escalation details. These details always help you to verify and ensure that promised functions are moving properly and expedite the requirement during the business urgency.

In addition to this, the same document can be digitized for access to the different departments on BYOD and Desktops in the Secure LAN.The Older Strategy of adding the Escalation and Penalty on the SLA can be standardized to Rewards to bring inspiration and overall improvement.

Requirements for SLA and AMC.

1. Comprehensive Site Survey and Installation report.

2. Timelines to regulate Prompt Delivery of Shipment.

3. Central Helpdesk to comply on the Technical Reports, Firmware updates, Remote Monitoring Tools.

4. Compliance to Demonstrate First Site Installation by Employed Certified Product Engineers and RCA Report on every failure.

5. Fault Isolation and Ticket closure within 48hrs to 72hrs.

6. Provision of Annual Certification from the Manufacturer stating as certified Vendor.

7. Downtime to support Regulation of Five nine.

8. Add Inspiration Clause to support Business continuity.

Business Partner Compliance parameters

- 24 x 7 HelpDesk Management.

- Online Ticket Monitoring System.

- Knowledge on using Remote Management Tools like Webex, Twitter, Logmein and GoTo Meeting.

- Ability to define Timelines with SLA.

- AMC contract with Principal Supplier.

- Contract copy with logistics Vendor.

- Service Delivery to have consistency with quality score.

- Quarterly Audit reports on the SLA Timelines and Service deliverables.

Best Practices for Small Business Owners

- Always Ask for the official meeting.
- Get Bank Account details of all the business partners (BP).
- MoA of the Company registered.
- MoU copy between the BP and other Associates.
- Visit Quality check department before signing SLA.
- Visit BP office every 3 months for the Training.
- Check physical revenue of last 6 months earned by other reference sites.

AMC Price Benchmarks

PRICING on Comprehensive AMC is available with Principal Vendor between 5% to 12% of the Billed Invoice. Repair TAT is between 7 days to 3 weeks including the shipment and logistics Management.

BUILDING UP NETWORK

It's an opportunity to experience the cultural community outside the walls of our own Network and our environment. Join an expat community that encompasses over the Personal Network. You will find like everyone is in the same boat—trying to network and connect with as many people as possible.

I suggest taking a leap of faith and leading your own initiatives. Build a supportive network at the earliest and there are numerous ways to do this, and I don't mean a database of cards or LinkedIn contacts.

Immerse yourself in the city's inner circles with purpose. Know your passion points and pursue events that reflect them in order to meet people that share your interests. Eventually, the people you meet will become the best references for the opportunities. Here's another tip you won't hear too often: try to stay behind at events that you love and offer the organizers help with future projects. Be a giver, not a taker and opportunities will fall into place

ANNEXURE : CASE STUDIES

I really feel happy to share you one of the unique experience about case study on the Project Development of HIGH SPEED PRINTER. Here this project was exercised at the Management Institute by using Simulation Engine.

But in real practical world with multi-culture platform the PM (project management) is all about the emotional challenge.

PM on Harvard Business Simulation Engine.
Project Score = 902/1150

Target Scope :
Product Design of new Multifunction Printer with WiFi, high speed Color scanner, copier and Fax module.

Team Process includes one to one coaching, Daily standup and Status review.

Project Scope has 200 task including the snap in module for scanning and faxing with more sophisticated features than expected competitors printer.

Project Objectives : Develop Multifunction Printer in 17 weeks at cost of $60,000. The Management will provide additional bonus points for completing task ahead of schedule, by keeping 85% team morale and under budget.

Decision Process allows to submit the plan every week with Project Resources like team size, skill level, amount of outsourcing, Weekly meetings and allowance with overtime.

Dashboard provides Graphical outcome of Team productivity, Team morale, Project Overview, Team process and Management Targets

Digital FCT/GSM Gateway Project Implementation

GSM PRI Gateway enables the route of calls from fixed Lines to mobile equipment by establishing a mobile to mobile call. In terms of savings GSM Gateway leverages business growth by 45%…. Here are the Facts used by Enterprise customers.

1) Almost 72% of the calls are terminated at the Mobile numbers.

2) Any call from the Land line is dialed with Pulse rate of 60 secs for Mobile numbers.

These 72% calls will have high revenue if dialed from the Land line supported with Pulse of 180 secs. On dialing from the SIM Interface to Mobile, the cost will come down by 55% thereby providing you with savings of 45%.

Advantages:

a) Higher efficiency for Domestic BPO companies.

b) Easy Integration to PRI Interface on PBX .

ROI is available from 12 to 18 months months, depending on the call volume.

The best way to test the product is to establish Multi party conference (preferably 5 to 6 party conference) and check for the time-slots or resources used by the Integrated Processor, MOS level for Voice quality and AMR codec.

GSM PRI Gateway uses EFR codec (Effective Full rate 12.2Kbps) or and AMR codec (Adaptable Multi Rate codec from 4.75kbps to 12.2Kbps). Some manufacturer provide GSM Gateway with Half AMR in terms of competition, but such gateway will run 60% capacity on full load.

MOS is the Mean opinion Score with the Toll quality Voice, where MOS=5 is for the Excellent Toll quality voice. The Multi party conference, is best way to test the Performance of Codec.

Managing GSM Gateway is 99% Easier by connecting the Product Remotely on Webex / Twitter / logmein.

WOULD YOU LIKE TO MAKE DENT IN THE UNIVERSE

There are no secret top-10 things you need to do to break out of the rat race. The most important concept to find your unique method is to understand yourself and define what you stand for remembering all the time that this construct itself can and should change as you mature. The second thing is the create an extremely robust learning environment for yourself that will hold good throughout life. This micro learning mind-set differentiates the you from others in the rat-race. The third thing is to believe that you are are destined for some unique achievements - the trick is to find that space where this can be done. Keep looking for the unique thing that you do and believe that you are designed to make an impact on the world. Once you have these basic building blocks - your attitude to life will change and the rest of the things that are essential will come naturally.

TECHNOLOGY KNOWN

1. IT Infrastructure (LAN/WAN, PoE, Server Rack Design, Cat5/6e, Wireless, Router, Switch, Firewall, GSM Gateway).

2. Fixed mobile Convergence on IP /GSM /WiFi /BTS Network.

3. Telecom Switch : Tadiran Flexicom, Siemens HiPath 8000, Alcatel, Nortel Meridian 11C, , Asterix/Centos.

4. Gateway   : FMC, GSM/PRI, T1/E1 –IP, Cisco Router 1721,

5. Data Analytics   : SQL /Oracle Hyperion with Crystal Reports.

6. Email Server   : MS Exchange 5.5, MDaemon, Lotus Domino.

7. IT Networks   : Enterasys LAN, SAN, Alcatel/Nortel WAN, GSM, BTS, IP Telephony, WiFi, VPC, SoftVPN,

8. Quality Certification   : ISO 9001 and HIPPA Standards for ITeS.

9. Art of Fulfillment   : Execution, Timeline, ROI, Gap Analysis.

10. Vendor Governance   : Due Diligence, Negotiation, SLA, MSA, NDA, Liaison, AMC.

11. Enterprise Dialer   : ASPECT, Touchstar with CTI /IVR /ACD/LCR IP Telephony and Voice Logger.

12. Enterprise System   : Alcatel MSC7270, Tejas /RAD Mux on IPLC, Nortel Passport 7440/7480, F5 WAN Accelerator.

13. Software Protocol   : LDAP, Java, OOP, Linux, MultiCast FTP, DHCP, DNS, SNMP, G729/G723, Qsig/SIP Trunk, H.323, IP SIP.

14. Network Security   : IPS, IDS, Firewalls, IPsec VPN, DMZ & Content filtering on Watchguard /Fortinet /CyberOam.

15. Servers /Platform   : Windows (NT/XP/2003), IIS, Linux Squid, Sun Solaris M9000, IBM /HP/Compaq, Web proxy.

16. Virtualization   : CITRIX Thin Client, VMWare

17. Storage & Backup : VERITAS, Tandberg LTO, EMC DMX3/VMax.

18. Telecom circuits : IPLC, VOIP, ISDN PRI, BRI, DS1, T1, DS3

19. Mainframe   : DEC 3800 M-Vax system with Thermal Recorder, Display monitor & M5 drive.

20. Power Staging   : UPS on SNMP alert with Redundancy.